BY V.C. THOMPSON

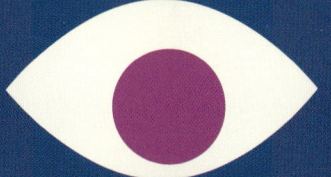

CONSPIRACY THEORIES
DEBUNKED

THE EARTH IS FLAT

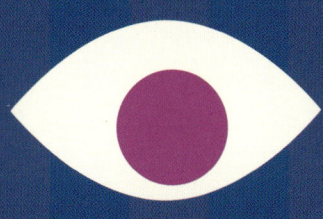

 45TH PARALLEL PRESS

Published in the United States of America by Cherry Lake Publishing

Ann Arbor, Michigan

www.cherrylakepublishing.com

Editorial Consultant: Dr. Virginia Loh-Hagan, EdD, Literacy, San Diego State University

Reading Adviser: Beth Walker Gambro, MS, Ed., Reading Consultant, Yorkville, IL

Photo Credits: © Jessica Rogner, cover, 1, interior graphics; © Rostislav Ageev/Shutterstock, 4; © George Stock/Wikimedia, 7; © Library of Congress/LOC no. 2011594831, 9; © MysticaLink/Shutterstock, 11; © James Jones Jr/Shutterstock, 13; © simonizt/Shutterstock, 14; © NASA/JPL/USGS/PIA00342, 17; © MISHELLA/Shutterstock, 21; © NASA/s65-20864, 22; © Woraphong Suphutayothin/Shutterstock, 25; © DC Studio/Shutterstock, 26; © AstroStar/Shutterstock, 27

Copyright © 2023 by Cherry Lake Publishing Group

All rights reserved. No part of this book may be reproduced or utilized in any form or by any means without written permission from the publisher.

45th Parallel Press is an imprint of Cherry Lake Publishing Group.

Library of Congress Cataloging-in-Publication Data has been filed and is available at catalog.loc.gov

ABOUT THE AUTHOR

V. C. Thompson is a nurse and author living in Michigan. V. C. is husband to his lovely wife, Anna. Together they are the proud parents of two children, one in Heaven and one newborn. His mission in life is to share his faith through the written word and the healing hand.

TABLE OF CONTENTS

INTRODUCTION 5
CHAPTER 1: THE PREMISE 8
CHAPTER 2: OUR VIEW 12
CHAPTER 3: FAKE PHOTOS 16
CHAPTER 4: FAMOUS ASTRONAUTS 20
CHAPTER 5: LASER POINTER 24
CHAPTER 6: THE VERDICT 29

TRY THIS! 30
LEARN MORE 30
GLOSSARY 31
INDEX 32

There are many unsolved mysteries. The Phaistos Disc is one. Some experts believe this ancient disc is a fake. Other experts believe it's real. Unfortunately, the disc cannot be examined. Examine means to study. The museum displaying the disc refuses all examinations.

4

INTRODUCTION

The world is filled with mysteries. Mysteries are things or events that cannot be easily explained. Some mysteries are conspiracies. Conspiracies are secret plans made by a group of people. Most conspiracies are stories that have no evidence. Evidence is proof. It is facts or information that supports a claim. Some conspiracies are later found out to be true. But most conspiracies are false. Sometimes it is hard to tell which conspiracies are true and which are false.

People who believe in conspiracies are called conspiracy theorists. Theorists are people who explain things with ideas called theories. Most people think conspiracy theorists are wrong. Some conspiracy theories are easy to prove wrong. Others are harder to prove wrong. Some are funny. Others are very serious.

Some people call conspiracy theorists crazy or stupid. This is hurtful. It is wrong to attack people, even if what they think is incorrect. Instead, the conspiracy theory should be examined. To examine a conspiracy theory, look at the evidence. Almost every claim has evidence. What matters is *how much* evidence there is and how strong it is. This tells us if the theory is true or false.

Let's take a look at a popular conspiracy theory. Don't forget to keep track of all the evidence. At the end, see if you can debunk this conspiracy yourself!

Sometimes conspiracy theories turn out to be true. UFOs and aliens used to be a conspiracy. Now, the government is investigating sightings.

CHAPTER 1:
THE PREMISE

Earth is 1 of 8 planets in our solar system. It is the third planet from the Sun. It's not too hot. And it's not too cold. But temperature is not the only reason why we can live here. In fact, there are countless reasons why there is life here. For instance, Earth has an ozone layer. This layer protects us from harmful sunlight. Earth also has water. Scientists have yet to find another planet where humans can survive. For now, Earth is our only home.

Earth is also the home of many mysteries. What is in the center? How is there so much water? Where did it come from? These are all important questions. You might have had the same questions. Have you also questioned Earth's shape?

For a long time, people thought Earth was flat. Many early ancient civilizations thought this. A few ancient Greek philosophers challenged that theory. A philosopher is someone who studies ideas like right and wrong. These particular philosophers lived about 2,500 years ago! They included Pythagoras (puh-THAH-guh-ruhs) and Eratosthenes (er-uh-TAHS-thuh-neez). Pythagoras is most famous for his discovery in math. Eratosthenes is famous for measuring Earth.

In the 1960s, people finally traveled to outer space. They proved Earth is a sphere to flat Earthers. A sphere is a ball shape. The astronauts took photos of Earth from space. So that settles it . . . or does it?

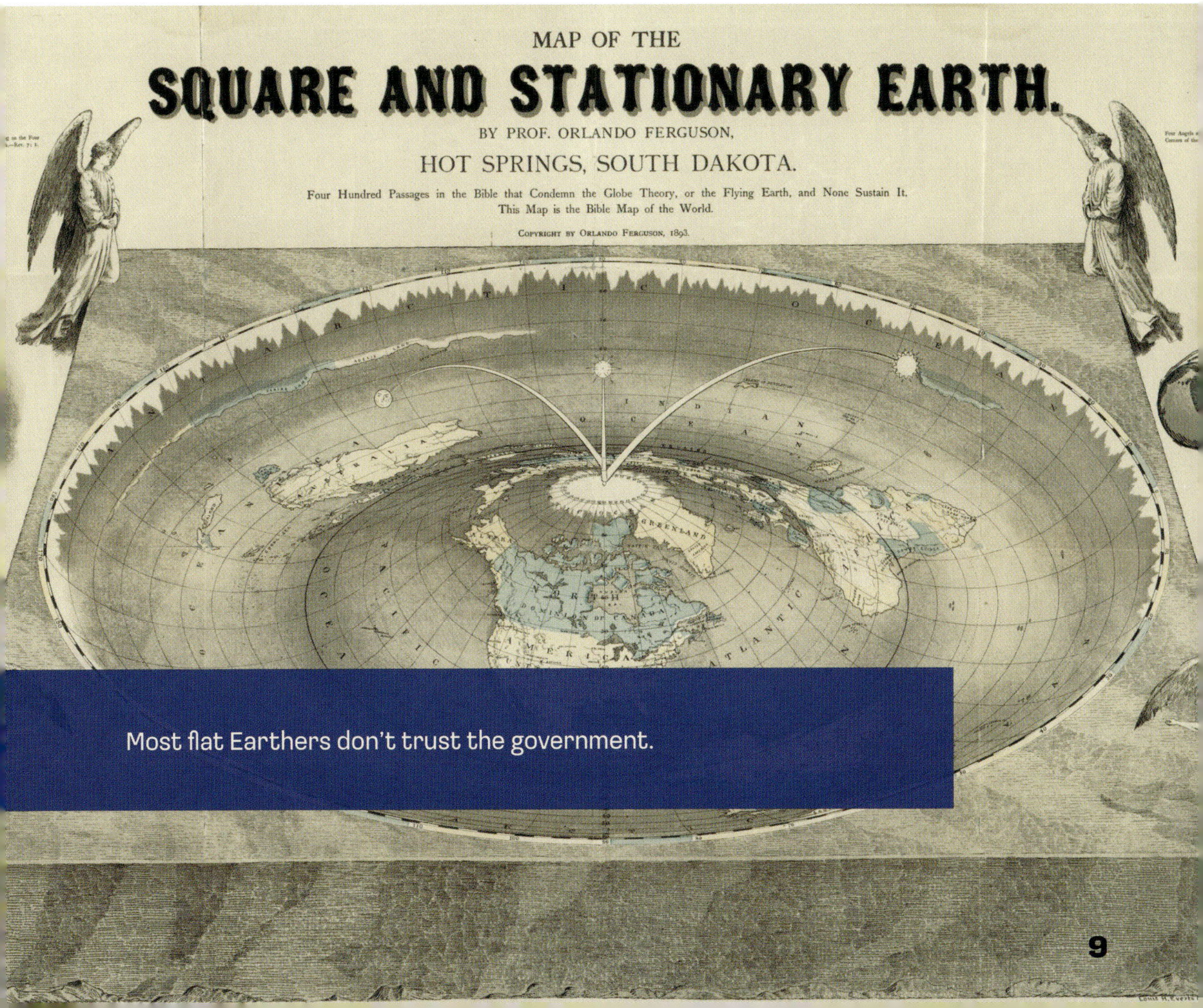

Most flat Earthers don't trust the government.

SPOTLIGHT

PYTHAGORAS

Pythagoras was an ancient Greek philosopher, mathematician, and astronomer. He also made contributions to music. He lived from 570 BCE to 490 BCE. He founded a secret religious group named after him. Its members were called the Pythagoreans. They didn't eat meat. They worshipped the Greek god Apollo. Pythagoras had great influence. He also inspired another famous philosopher, Plato.

Pythagoras thought Earth was round. He was the first to propose this theory. Greek philosophers such as Pythagoras tracked the movement of stars and the Moon. From this, Pythagoras realized Earth was round. He also reasoned that because the Moon is round, Earth must be too.

Some people don't believe Earth is a sphere. Instead, they believe it is flat, like a disk. This is called the flat Earth conspiracy theory. They say that NASA has been lying to us. NASA is the National Aeronautics and Space Administration. It is part of the American government. NASA workers are scientists and astronauts who study space. The conspiracy theorists believe the government is bad. They think NASA faked Earth photos to fool us. They even think that the Moon landing was faked. Some even say it is impossible to go to space. What do you think? Are they tricking us? Is Earth actually flat? Let's take a look.

CHAPTER 2:
OUR VIEW

THE CONSPIRACY

Have you ever been in a tall building? Have you ever stood in a large, flat field? Have you ever been to the beach? What did you see there? Did you notice the horizon? A horizon is where we see the sky meet Earth. It is far away. The horizon is always a straight line. This is evidence that Earth is flat. The horizon would also be curved if Earth were a ball. Right?

We see the Sun set and rise at the horizon.

Earth is massive compared to humans. When we stand outside, Earth may appear flat.

DEBUNKED

It is true that horizons are flat. Or, at least, they look flat to us. The horizon does curve. The curve is too small to see at our height. According to scientists, Earth curves around 8 inches (20 centimeters) per 1 mile (1.6 kilometers). Can you see a curve of 8 inches (20 cm) from 1 mile (1.6 km) away? Probably not.

However, the higher you go, the more likely you'll see Earth's curve. In fact, the curve can be seen at around 6.6 miles (10.6 km) high. That's pretty high. Lucky for us, people have flown that high. The recordings from spaceships show the horizon bending slightly. The horizon gets more bent as the camera gets higher. Eventually, you can see the whole sphere of Earth!

CHAPTER 3:
FAKE PHOTOS

THE CONSPIRACY

You've probably seen pictures of Earth. These are pictures taken from outer space. Many were taken by astronauts on the Moon. Some people believe these photos are all fake. If you look closely, the continents change size. For instance, look at North America. It was small in 2002. Then in 2012, it was big. In 2015, the continent was small again! How did an entire continent grow and shrink? NASA must be using fake photos.

Look at the NASA photos of Earth from 2002 and 2012. Look at photos side by side. Notice the clouds. Some parts of the clouds look identical to each other! How can this be? After all, no 2 clouds are identical. It looks like NASA Photoshopped the images. Photoshop is a computer software program that makes changes in photos. Is NASA faking the pictures?

NASA admits to editing images. But it's for a good reason! NASA says editing the images helps others better see space.

SPOT-

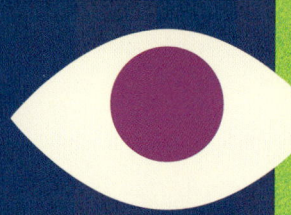

PSYCHOLOGY

Our brains are very powerful. Our eyes are too. But sometimes they trick each other. For example, the Moon is a sphere like Earth. A full Moon is when the entire Moon is bright. It looks like it is making light like the Sun. But don't let your eyes fool you! The Moon makes no light. It is simply reflecting light from the Sun back to us.

In fact, you can see this in action. The Moon circles Earth. As the Moon moves, it goes from new to full. A new Moon is all dark. A full Moon is all lit up. It takes the Moon almost 30 days to complete its cycle. This is called the lunar cycle. The dark part of the Moon is the part the Sun doesn't shine on. Things are not always as they first appear! Conspiracy theories are often the same way. Conspiracy theories may seem to make sense. But after a little more digging, they are usually debunked.

LIGHT

DEBUNKED

At first, Earth seems different in NASA's photos. There are 2 explanations for this. The first explanation is that the object could be different. The second explanation is that the perspective could be different. Perspective means view.

In this case, the perspective is different. NASA took these photos at different distances. At a closer distance, the camera sees less of Earth's surface. Farther away, it sees more. Up close, the continent looks bigger. Farther away, the continent looks smaller.

Now let's look at the clouds. They look Photoshopped. That is because they are! NASA made the photo from many pictures. A camera in space circled Earth for 2 days. During this time, it took many photos. Each photo only showed parts of Earth. NASA combined several photos to create 1 photo. Then they made it look better with Photoshop. This is why there are duplicate clouds.

CHAPTER 4:
FAMOUS ASTRONAUTS

THE CONSPIRACY

Mirosław Hermaszewski was a famous astronaut from Poland. He flew to space in 1978. In space, he saw Earth. Years later, in 2018, he was interviewed on television. He was asked if Earth was flat. He answered that in space he saw a flat Earth. Hermaszewski broke the secret! He saw the flat Earth for himself and told everyone. This is eyewitness proof! Do you believe him?

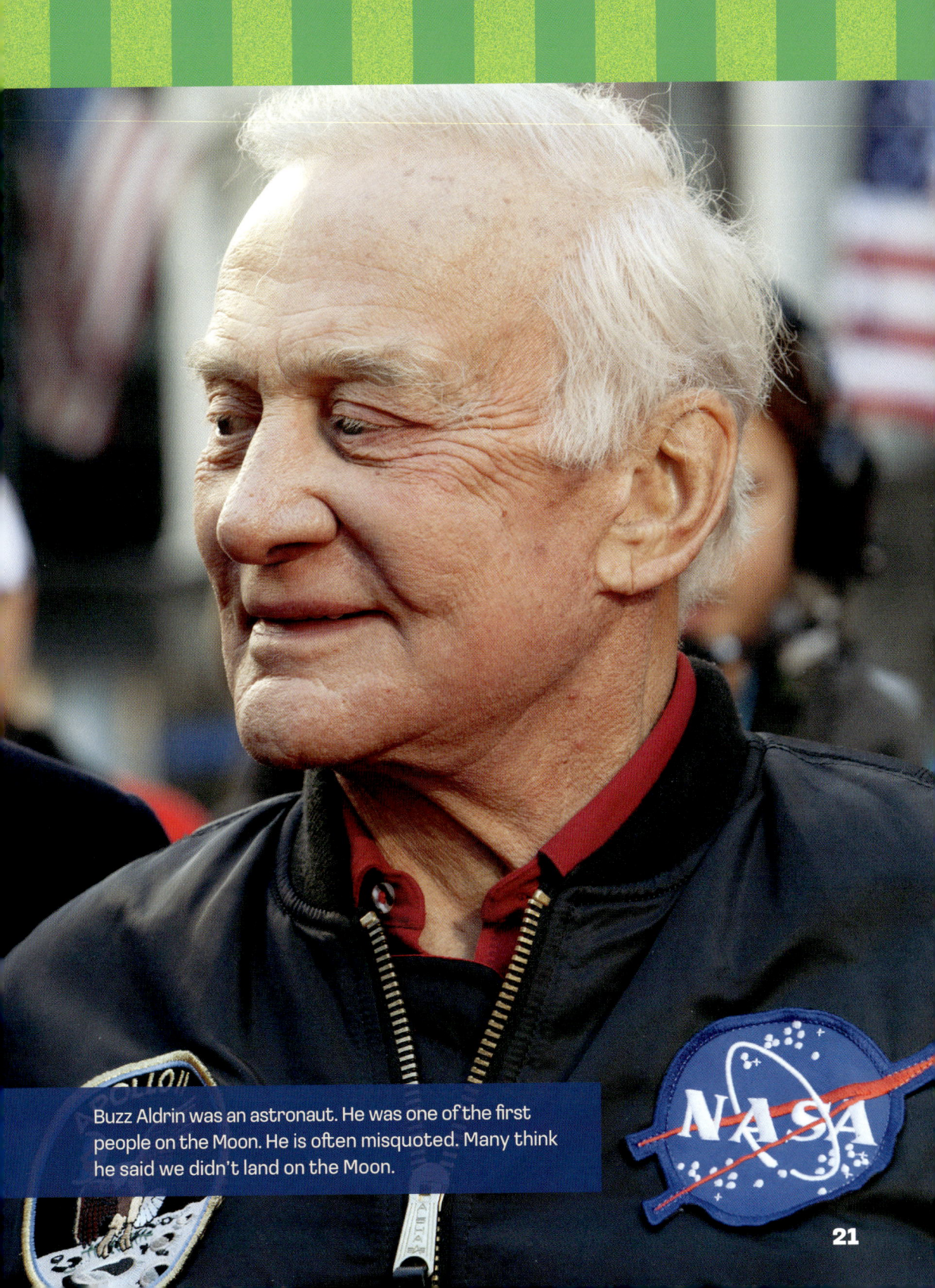

Buzz Aldrin was an astronaut. He was one of the first people on the Moon. He is often misquoted. Many think he said we didn't land on the Moon.

Sometimes statements are taken out of context. They can be misinterpreted. Sometimes this is done on accident. Other times, it is done on purpose.

22

DEBUNKED

Mirosław Hermaszewski really did say Earth is flat. But what was the context? Context is the full situation around something. Context helps us know what is going on. Say you are acting in the school play. On opening night, your friend Zoe tells you to "break a leg." That's a mean thing to say, right? Not if you are about to go on stage. In theater, this means "good luck." Going on stage in theater is the context.

It is the same for Hermaszewski. What was the context of the interview? The interviewer was asking funny questions. So Hermaszewski was giving funny answers. In the same interview, they asked if he saw aliens. He joked that he could not say while on camera. It seems Hermaszewski is not really a flat Earth believer.

But let's suppose Hermaszewski was being serious. Hundreds of people have been to space. They all say Earth is a sphere. What is easier to believe? One man lying about Earth being flat? Or hundreds of people lying about Earth being a sphere? It is more likely that just 1 man would lie.

23

CHAPTER 5:
LASER POINTER

THE CONSPIRACY

Some flat Earthers have experimented with lasers. Lasers are straight lines of light. Lasers travel in 1 direction until they hit something. If Earth is round, the ground should curve. Scientists say that the ground curves down 8 inches (20 cm) per 1 mile (1.6 km). But when people shine lasers across 1 mile (1.6 km), it is at the exact same height. This proves there is no curvature. This means Earth is flat, right?

A laser's light will continue in a straight line. This will happen if there is no curve.

25

Rigging a science experiment is unethical. Unethical means against the standards of a profession.

DEBUNKED

These laser pointer experiments are tricky. They seem convincing. But the flat Earthers have tricks up their sleeves.

It certainly looks like the laser stays at the same height. This is because the experimenters angle the laser slightly up. They might not be doing this on purpose. But when this happens, the laser might hit 8 inches (20 cm) higher than normal. This is rigging the experiment. Rigging is cheating in a situation to get what you want. Rigging is not allowed in science experiments.

SPOT-

SCIENCE EXPERIMENT

Scientists are always experimenting. A good experiment includes observations, a *hypothesis*, independent and dependent *variables*, and results.

Let's say you get goose bumps when you are cold. The observation is you get goose bumps when it's cold. You notice your goose bumps when it's 65 degrees Fahrenheit (18 degrees Celsius). You think you start getting goose bumps when it's 65°F (18°C). This statement is your hypothesis. You develop an experiment.

You sit in a room wearing a T-shirt and jeans. You have your friend lower the temperature 1 degree every 5 minutes. At 67°F (19°C), you get goose bumps. The independent variable is the temperature. This is what's being adjusted. The dependent variable is whether you get goose bumps. Now you know the answer to your question!

Scientific testing can also be used to test conspiracy theories. What would the observations, hypothesis, independent, and dependent variables test the flat Earth theory?

CHAPTER 6:
THE VERDICT

Flat Earth theories seem believable. The evidence seems convincing at first. How is Earth round if I can't see the curve? Why do lasers stay at the same height for miles?

However, we now have good explanations for all these questions. Sometimes you can't trust your eyes. Perspective matters. Knowing the whole story is important. Context matters. Not all experiments are free from error.

Let's finish by asking a few questions. If Earth is really flat, why would NASA lie to us? Why don't sailors and pilots fall off the edge? Why wouldn't the whole world have night at the same time?

These are all hard questions for these conspiracy theories. It is good for both sides to ask questions. Whatever the shape, Earth is home to us all. It is also home to many mysteries. But the flat Earth theory isn't one. This conspiracy theory has been DEBUNKED. Or has it? What do you think?

TRY THIS!

1. Find a ball. Pretend it is the Moon. Then find a flashlight. Pretend that is the Sun. Turn off the lights. Then have a friend shine the light on the ball. Move the ball around. Can you see the new Moon, half Moon, and full Moon?

2. Take 2 pictures of a globe. Take the first picture from a distance of 7 inches (18 cm). Take another picture from 10 inches (25 cm) away. Do the continents look like they shrank?

3. With an adult's permission, experiment with a laser pointer. Be sure not to shine it in your eyes or those of others! How far do you think the laser can travel?

4. Form a scientific theory. Then test it with an experiment. Was your theory right?

LEARN MORE

Goldstein, Margaret J. *What Are Conspiracy Theories?* Minneapolis, MN: Lerner Publications, 2020.

Johnson, Anna Maria. *Debunking Conspiracy Theories.* New York, NY: Cavendish Square Publishing, LLC, 2019.

GLOSSARY

conspiracies (kuhn-SPIHR-uh-seez) secret plans to do something bad or against the law

context (KAHN-tekst) information surrounding a word or group of words

evidence (EH-vuh-duhnss) facts or information that supports a claim

horizon (huh-RY-zuhn) where the sky meets Earth

hypothesis (hye-PAH-thuh-suhs) educated guess or possible explanation

lasers (LAY-zuhrs) straight beams of light

lunar cycle (LOO-nuhr SY-kuhl) transition from new Moon to full Moon

mysteries (MIH-stuh-reez) things that are unknown or hard to explain

perspective (puhr-SPEK-tiv) point of view

philosophers (fuh-LAH-suh-fuhrz) people who study

Photoshopped (FOH-toh-shahpt) used Photoshop software to change a computer photo

rigging (RIH-ging) cheating in a situation to get what you want

sphere (SFEER) ball shape

theorists (THEE-uh-rists) people who explain things with ideas called theories

variables (VER-ee-uh-buhls) factors in an experiment that may be subject to change

INDEX

Aldrin, Buzz, 20
aliens, 7, 23
ancient philosophers, 8, 10
astronauts, 11, 20-23
astronomers, 10, 11

clouds, 16, 17, 19
conspiracies, 5, 18
   believers and
      communication, 6
   testing and questioning,
      28, 29, 30
conspiracy theorists,
   6, 11, 24, 27
context, 23
curvature of the Earth,
   12, 15, 24

debunking and disproving,
   6, 18, 29
   astronauts theory, 23
   fake photos theory, 19
   horizon theory, 15
   laser pointers theory, 27

Earth
   curvature of, 12, 15, 24
   flatness conspiracy
      theory, 11, 12-15, 20,
      23, 24, 29
   horizon of, 12-13
   life conditions, 8
   photography, 11, 15,
      16-17, 19
   scientific reasoning,
      8, 10, 29
   size, 14, 15

Eratosthenes, 8
evidence, 5
   doubt of photography, 11,
      16-17, 19
   flat Earth theory, 12, 29
   supporting or debunking
      conspiracies, 5, 6, 28, 29
experiments, 26, 27, 28, 30
eyewitness accounts, 20, 23

fake photos, beliefs, 11,
   16-17, 19

government distrust, 9, 11
government investigations, 7
Greek philosophers, 8, 10

Hermaszewski, Mirosław,
   20, 23
horizon, 12-15
humor, 23

interviews and quotations,
   20-23

laser pointers, 24-25, 27, 30
lunar cycle, 18

Moon, 10, 18, 21, 30
Moon landing, 11, 21
mysteries, 5, 8

NASA (National Aeronautics
   and Space Administration)
   photography, 16-17, 19
   space travel, 11, 21, 22

personal attacks, 6
perspective, 19
Phaistos Disc, 4
philosophers, 8, 10
photography
   doubt of evidence, 11,
      16-17, 19
   Earth's curvature, 9, 15
   editing, 16, 17, 19
   experiments, 30
Photoshop, 16, 19
planets, 8
Plato, 10
Pythagoras, 8, 10

quotations and interviews,
   20-23

rigging of experiments,
   26-27

science experiments,
   26, 27, 28, 30
solar system, 8
space travel
   Earth photography,
      9, 15, 16-17, 19
   history, 9, 11, 20
symbols, 4

UFOs, 7